IN THE LAND OF COTTON

In the Land of Cotton

How Old Times There Still Shape Alabama's Future

Larry Lee

NewSouth Books

Montgomery

NewSouth Books
105 S. Court Street
Montgomery, AL 36104

Publisher's Cataloging-in-Publication data

Lee, Larry.
In the land of cotton : how old times there still shape Alabama's future /
Larry Lee.
p. cm.

ISBN 978-1-60306-397-5 (paperback)
ISBN 978-1-60306-398-2 (ebook)

1. Cotton trade—Southern States—History. 2. Alabama—History.
I. Title.

2015936104

Cover photographs:
Front, top left, Kimberly Vardeman (Flickr: Cotton Harvest).
Front, top right, Tom Murphy VII. A cotton gin on display at the
Eli Whitney Museum, Hamden, CT.
All other images courtesy of the author.

Printed in the United States of America

I.

Very few contemporary Alabama schoolkids have ever seen a stalk of cotton up close; none know what it is like to drag a pick sack down long rows as they pluck the white fiber from the thorny bolls. Yet our once pervasive "cotton culture" still leaves its imprint on who and what we are today, coloring everything from how we view the importance of education and fund it to our insistence that low taxes and cheap labor and multiple economic development incentives will get us to the industrial promised land.

How we got to this point is a complicated—and sometimes tragic—story, but understanding where we've been is critical to understanding where we can go.

Cotton was the lifeblood of rural Alabama for a century and a half. The relationship started even before Alabama became a state in 1819. Two things set the love affair in motion. The first was the invention of the cotton gin by Eli Whitney in 1793. The second was the defeat of the Creek Indians at Horseshoe Bend by Andrew

Jackson in 1814.

Whitney's invention paved the way for significant increases in cotton acreage while Jackson's victory opened Alabama to settlers from Tennessee, Georgia, the Carolinas, and Virginia. With "Alabama Fever" in full bloom, the state's population increased more than 1,000 percent from 1810 to 1820.

By 1839, half of all U.S. cotton exports went through the port of Mobile. By 1840, Alabama had 253,532 slaves. In 1850, we produced 564,000 bales of cotton.

However, the truth is that most of Alabama never fit the stereotype of an endless landscape of antebellum mansions, horse-drawn carriages, and legions of docile servants. In 1860 only 1.3 percent of the state's 50,064 farms had more than 1,000 acres, while 68 percent were less than 100 acres.

My family was typical of the yeomen farmers spread across Alabama. My great-great-great grandfather, William Greenberry Lee, left Putnam County, Georgia in 1823, traveled across the Indian territory of east Alabama, and settled in south Butler County. As daddy used to say, "He was poor when he left Georgia and when he got to Alabama, he stayed that way."

After the Civil War, slavery was replaced by tenantry and sharecropping, but little else changed. For almost another century, many rural Alabamians—white and black—planted cotton in the spring, chopped it in the summer, and picked it by hand in the fall. For decades, the future for most was no farther than getting to the end of the next cotton row or putting the mule in the barn at sunset. The children of sharecroppers were far more likely to hear the rasp of a pick sack being dragged on sandy soil than the ringing of a school bell.

The cotton culture did offer one alternative: the cotton mill, though many who worked in them simply traded their hardscrabble life on the farm for an equally subsistence existence in the mill village.

Our first cotton mill was built near Huntsville on the Flint River around 1820. By 1832, the Bell Factory, with 3,000 spindles and 100 looms, was operating in Madison County.

In his book, *Cotton and Race in the Making of America,* Gene Dattel states that "In many ways, cotton had been the oil of the 19th century." Alabama did its part to see that this was the case, not only in the 19th century, but through at least the first half of the 20th.

But not without paying a price.

During the post-Reconstruction period, Atlanta newspaperman Henry Grady proclaimed a "New South" and Southern leaders worked feverishly to make Grady's dream a reality. For the most part this meant we looked toward New England and chanted "cheap labor, cheap land, low taxes."

In 1897 the state of Alabama passed a law exempting anyone who invested $50,000 in a textile mill from all state, county, and municipal taxes for 10 years. In 1899, Alabama produced $22 million worth of cotton goods, ranking us ninth in the country. By 1900 we had 31 textile mills. That same year the average annual wage of a male working in manufacturing in Alabama was $309, while his counterpart in Massachusetts made $527.

And in Elmore, Chambers and Talladega counties, more than 20 percent of the manufacturing labor force was children.

As Wayne Flynt states in *Alabama in the Twentieth Century,* "Alabama became one of the four leading textile states. But the sector of that industry that relocated to the South from New England was the least profitable, used the least skilled labor, and thrived on a family wage system that required women and children to work in order to provide families a bare living."

Things hadn't changed much by 1930 when my granddaddy was one of Alabama's 166,000 tenant farmers.

President Franklin Roosevelt described the South as "the nation's number one economic problem" in 1938. This statement was met with righteous indignation across the region. Yet the Alabama State Planning Commission published a booklet, *Industrial Opportunity in Alabama,* in December 1944, which stated: "At present Alabama possesses what may be termed low-grade industries; that is, industries which require considerable unskilled labor, and which pays low wages."

The writer of this passage was astute because "unskilled" is simply a synonym for not well educated. There is comfort in continuity, but change is threatening. We did not diversify our economy—or our children by stressing to them the value of education.

To understand the extent of this culture, let's look at 1950. Cotton was grown in every county, a total of 1.3 million acres. Covington County had 18,500 acres, some of them on my grandfather's farm near Red Level. And every county except Bibb had textile mills or apparel plants. There were more women (1,274) running sewing machines in Covington County than any other Alabama county. Some 67 percent of all females with manufacturing jobs in the county worked in apparel. In the state's Piedmont region, cotton mills dominated the economy. Nearly 90 percent of all manufacturing jobs in Chambers County were in a mill. In neighboring Tallapoosa County the figure was 78 percent.

In 1950, Harry Truman was in the White House and I was in the first grade at Church Street School in Andalusia. Every day at noon the whistle blew at the "shirt factory" eight blocks away. While the whistle told workers it was lunch time, in reality it was telling the world that Alabama's cotton culture was alive and well. Looking back over six decades, I know the whistle could also have been a lamentation about the vise-like grip cotton had on the state, a grip that still lingers.

In 1950 the good citizens of Andalusia thought the factory whistle would sound until Gabriel blew his horn. But it fell silent 20 years ago. Across Alabama today, buildings where workers once breathed cotton dust and risked arms and hands stand empty except for the occasional school boy hurling rocks to break out another window. And low-slung buildings where sewing machines once whirred watch as kudzu creeps across vacant parking lots.

In October 2014, the 11 counties with Alabama's highest unemployment rates are all rural; these same counties in 1950 had 3,317 textile jobs and cultivated 182,400 acres of cotton.

Dr. James Cobb is the Spaulding Professor of History at the University of Georgia and has written extensively about the South's effort to improve its economy. Here's how he describes what went on in rural areas for decades.

> . . .we might take a quick glance at the places where, a half-century or more ago, local leaders had decided to mortgage their town's social and institutional future by wooing footloose northern industries with promises of cheap labor, construction subsidies, tax exemptions and guarantees of protection from unions and higher wage competition.
>
> These days, a great many—probably most—such communities have long since bade goodbye to their one-time industrial benefactors who skipped town in a hurry once they heard about the even warmer hospitality awaiting them in places like Honduras or Bangladesh. In the wake of their departures, meanwhile, their former hosts are enjoying little success in bringing in new employers for relatively high-wage (by global standards) labor with only low-wage skills. Such are the fruits of trying to achieve a developed economy at the expense of a developed society.

Looking back now I realize when I was at Church Street School that our pride in our wage differences and our insistence to give incentives to companies when we were only spending 56 percent per pupil of the national average to provide my schooling was sowing the seeds of a crop that rural Alabama has been harvesting for the last four decades.

Daddy spent his time in World War II glued to a radar screen helping airplanes land on the Azores Islands in the middle of the Atlantic Ocean. He stayed in the Air Force after the war, but hundreds of thousands of his comrades came home and made use of the G.I. Bill to go to college; the rate of college enrollment in Alabama went up 92 percent from 1940 to 1950.

And as the first Baby Boomers finished high school the rate of college enrollment jumped 100 percent from 1960 to 1970.

This is when our blind allegiance to the cotton culture began to subtly impact the future of rural Alabama in a way no one had ever considered. It was as if we were the rat and put the cheese in our own trap.

II.

To learn more about what has happened across the state, I questioned more than 200 people who grew up in 45 rural counties and graduated from high school anywhere from 1956 to 1993. The average graduation year was 1972.

Their average class size was 67. About 37 percent of each class went on to college, many to a two-year school. Interestingly, about 36 percent of those questioned had a parent who had been to college. However, many came from families where neither parent had a high school education.

Of those who got a college degree, less than 20 percent returned to the communities where they grew up. That works out to four college degrees returning while 14 did not, for each class of 67.

Those who came back basically had a family business or farm to return to, became a teacher, or became a doctor, attorney or dentist and came back to open a practice. And if they were a minority, they only returned to be teachers.

Why did the others not return?

"Lack of opportunity" was the answer I got over and over.

There were 18 members of the class of 1978 in a Chambers County school. One became an attorney and moved to Maine. One became a civil engineer and moved to Texas. One became a nurse and moved to Ohio. One went into the military. Four became teachers and stayed in the county. One graduated in agricultural science and moved to Birmingham. Of those who did not go to college, only one left the area.

One respondent who had grown up in a mill village in Tallapoosa County and said that many of his classmates who returned home had majored in textile management. "They thought they had lifetime jobs," he says wryly.

Ten of the 45 members of the class of 1962 of a Washington County school went to college. Three got their doctorates. None live in Washington County. A 1962 graduate from a Butler County school said: "Of those leaving home for college, the returnee is the exception. There are simply no local jobs for professional folk."

Fourteen of a Dekalb County class of 56 went to college. Only three, all teachers, are still in the county. Of a 1968 class of 62 in Lamar County, perhaps 15 finished college. Two teachers, a social worker and an engineer are still in the county.

Of course it would be foolhardy to imply that one cannot succeed without a college degree—Bill Gates being a case in point. Many of those questioned had parents who, while only high school graduates or less, had successful careers. Still, these mamas and daddies saw the value in education and encouraged their children to go beyond high school. They saw a day when it would be more important to be able to think than to be able to sweat.

So for decades now, the majority of our best and brightest young people across rural Alabama have been leaving home and coming back only for high school reunions, family holidays, and funerals.

A 1967 graduate of a Winston County school tells a story that sums up the situation well. "When I was in the 7th or 8th grade the principal came in our room and told us, 'All of you who finish high school will leave Winston County and not return. Those of you who drop out of school will stay in Winston County and your children will come to school and want to do the same thing. There is no hope for Winston County.'"

Any cattleman knows that you improve the quality of your herd by keeping your best heifers. Unfortunately, as the Winston County principal knew all too well, rural Alabama has been selling its best heifers for a long time.

Dr. Don Bogie, the former head of the Center for Demographic Research at Auburn University Montgomery looked at the rate of college enrollment per 1,000 persons 18 or older for 1970 and 2000. He studied four rural counties (Cherokee, Choctaw, Franklin and Geneva) and three urban counties (Houston, Montgomery, and Morgan).

It is hardly surprising that rural counties trail urban counties significantly in this measurement. Franklin had the highest college enrollment rate of the rural counties, but this rate significantly trailed that of urban counties.

Data from the Alabama Commission on Higher Education shows that in the fall of 2013, in 7 of the 11 high-employment counties mentioned previously, more than 40 percent of students who went on to college had to have remedial math or English. It was 69 percent in one high school.

So not only are rural students going to college at a much lower rate than those from urban areas, but those who do enroll lag behind in how well they are prepared.

School dropouts have always been a rural problem. Data from the Alabama Department of Education shows that over the last five years rural schools have had 30 percent of our public school students—but 36 percent of our dropouts. As long as the mills and sewing factories were in operation, dropouts had a place to go. This is no longer the case.

Ask any rural high school guidance counselor why kids today drop out of school and you will quickly be told, "Because they are not getting any support at home." But if mama or daddy dropped out themselves, is this surprising?

III.

Well into the second decade of the 21st century, rural Alabama finds itself in a dilemma, one that has not gone unnoticed nationally.

It has been noted that Alabama leads the nation in percentage of rural jobs lost since the recession began in December 2007. And a study by the Rural School and Community Trust determined that Alabama's rural schools need more attention than those of any other state in the nation.

It comes down to this: at a time in history when the race will go to those with the best schooling, rural educators across the state (already struggling because of economic constrictions) are trying to sell the importance of education to a populace that increasingly has never shown that they valued it.

The importance of good schools can be vividly illustrated by looking at the history of two nearly identical rural counties. Both have had an interstate for 40 years. Both are within 75 miles of major airports and metro areas.

In 1950, though one had 10,000 fewer people, it had more high school and college graduates than the other. One was still largely dependent upon agriculture while 50 percent of the labor force in the more affluent one worked in manufacturing. In median family income and percent of the population with high school degrees, one county ranked near the top of the state, while the other ranked closer to the bottom.

But over the next 50 years, their fortunes slowly but steadily began to reverse. In 1975, the lines tracking percent of high school graduates crossed, and by 1995 so did median family income.

By 2000, the affluent county of 50 years earlier was smaller than in 1950, while the other county had grown by 57 percent. Even more impressive is that in one county the number of high school graduates went up three times faster and college graduates four times faster than in the county that had more of each in 1950.

Two things happened: One county worked hard to diversify its economy while the other clung to its cotton economy. One understood that new companies need a well-educated work force while there is little evidence that the other did.

A look at test scores for the past five years for both county school systems is revealing. In one, all students in grades 3–8 scored above the state average in reading and math at level 4. In the other, the same scores were significantly below the state average.

So what is the lesson to be learned? First and most obvious is that whatever we've called "rural development" for decades has failed us miserably.

More than one million people call rural Alabama home. They deserve more than the lip service they've been given. We've talked way too much about "initiatives" and done way too little to put meaningful processes into place. We've held "forums" and "listening sessions" and created "commissions." We've measured "success" in how many people attended meetings, not in whether lives were being improved. We've looked for the "quick fix" so that we could pat ourselves on the back by the next election and brag about "progress" while ignoring empty factories, vacant homes and fewer students in classrooms.

But wringing our hands and reliving our failures is of no value whatsoever—unless we vow to truly understand what happened and pledge to devote our resources to reasonable and meaningful responses. Nor can we set timetables based on the next election. Rural Alabama did not come to this point overnight, and as distasteful as it may be for political leaders to accept, its fortunes will not reverse overnight.

The Rural Medical Scholars program at the University of Alabama is a great example of staying the course. Started in 1996 to encourage students from rural areas to get a medical degree and practice in a rural location, it took more than a decade for the program to return dividends. But given the fact that bringing a doctor to a rural community has the economic impact of $1 million or more annually, RMSP has contributed over $100 million to rural economies in the last six-seven years.

The two most pressing needs of rural Alabama involve education and the economy. When building a house you first put the foundation in place. Education and the economy must be rural Alabama's foundation.

As the Center for Rural Alabama pointed out in its 2009 study, *Lessons Learned from Rural Schools*, there are outstanding schools in every corner of Alabama. However, the outstanding ones are greatly outnumbered by those that at best are only average.

Several things must be done to improve rural education. One is to give rural principals every opportunity to improve their skills by offering professional development specifically designed for them. "One size does not fit all" is as applicable to helping rural school administrators as it is to helping rural economies.

We must redouble our efforts to engage communities in the education process. It is imperative that local citizens understand that education is vital to community survival. As has been pointed out earlier, this task will not be easy because of the steady erosion of the "education foundation" in most rural places. The faith-based community should play a key role in this because they remain an integral part of the rural landscape.

We must work hard to "grow our own" rural educators. Teachers and administrators who come from rural communities are most likely to remain and thrive in such environments. Yet the pipeline of such people is little more than a trickle.

The Ozarks Teacher Corps started in Missouri in 2010. College students from rural southwest Missouri who are planning to become teachers are awarded scholarships in return for committing to teach in a rural school location for at least three years. Alabama should look at a similar program.

There are no "silver bullets" that can solve education challenges overnight. However, political leaders seldom understand this and, consequently, we end up with legislation like the Alabama Accountability Act that is touted as helping rural schools when in fact it does the opposite.

Using rural students as the rope in a political tug of war is terrible and education leadership should not be timid in pushing back against those who promote bad policies.

IV.

Obviously it is impossible to maintain any semblance of

community without economic activity. But rural Alabama and statewide must look beyond our traditional manufacturing economy.

Grandpa would've scratched his head at the notion that housewives in affluent suburbs would pay a premium for "free-range" eggs or organically grown vegetables and meat or that "city folks" would spend their vacation working on a farm. Yet economic development is about bringing outside dollars to a community, whether in wages from a company, in purchases from ladies in Birmingham and Mobile who want brown eggs, in spending by outdoorsmen, in the income of retirees escaping the crime and congestion of a metro area, or in activity created by a new doctor in town.

A recent report points out that retiring Baby Boomers are far more likely to re-locate to a rural area than any other population segment. How does rural Alabama capitalize on this?

We need to devote far more energy and resources to helping entrepreneurs get their feet on the ground and to helping small business grow and succeed. Alabama has known for many years that about 75 percent of all new jobs annually come from industry expansions.

When a cabinetmaker hires one or two people, there is no groundbreaking or photo op for mayors, county commissioners and state officials. Yet we must recognize that by and large, rural Alabama is the land of small businesses. Census data shows that of the 25,000 businesses in rural communities, 96 percent have 49 or fewer employees.

State officials must understand that what may be small in one area, may be large in another.

For example, a legislator friend represents two rural counties and part of an adjacent metro county. Several years ago a company in one of the rural counties wanted to expand and add 50 jobs. He went to Montgomery to see what kind of assistance might be possible for the company and was told, "Fifty jobs aren't in our matrix."

Here's what Montgomery didn't understand: For every person in the labor force of the rural county in question, there are 12.8 in the metro labor force next door. When you multiply 50 jobs times 12.8 you find that the impact on the rural county's labor force was the same as 640 jobs in the metro location. No doubt had my friend told Montgomery that he was dealing with 640 jobs, he would've received a different response.

It's been decades since I picked cotton. It was hot, dirty hard work. Thinking back to those days does not make me wish for them again.

But doubt it not, we have yet to shed the rhythms and values Alabama's cotton culture embedded in our soul. However, Church Street School is closed. The factory whistle at the shirt factory no longer signals time for lunch.

It's way, way past time that we understand this.

About the Author

Larry Lee spent 45 years "studying country folks and country places." He directed the Center for Rural Alabama and is co-author of *Beyond the Interstate: The Crisis in Rural Alabama*; *Crossroads and Connections: Strategies for Rural Alabama*; and *Lessons Learned from Rural Schools.* He also directed the West Alabama Economic Development Authority, the Covington County Economic Development Commission and the Southeast Alabama Regional Planning & Development Commission. He frequently writes about education and is chairman of the advisory board of HIPPY Alabama, an early childhood learning program. An authority on rural development, Lee has presented to the Southern Legislative Conference, the Southern Growth Policies Board, and the Delta Regional Authority, as well as VOICES for Alabama's Children, Alabama Association of School Boards, Alabama Farmers Federation, Association of County Commissions, and School Superintendents of Alabama. Lee is a graduate of Auburn University and was an editor with *Progressive Farmer* before diving into community development.